HABITABLE EXOPLANETS
Red Dwarf Systems Like TRAPPIST-1

A Habitable Planets Series

Copyright © April 2018 Mathew Anderson
All Rights Reserved.

eBook published by Amazon:
www.amazon.com/author/mathewanderson

Follow on Twitter: @OurCosmicStory

No part of this book may be used, reproduced, or transmitted in any manner without written permission of the writer, except in the case of brief quotations in critical articles and reviews. For information, please write: ourcosmicstory@gmail.com.

Mathew C. Anderson

For All Great Explorers

"Equipped with his five senses, man explores the Universe around him and calls the adventure Science." – Edwin Hubble

ACKNOWLEDGMENTS

The motivation to start writing *Habitable Exoplanets: Red Dwarf Systems Like TRAPPIST-1* came partly from research accrued when I was writing *Our Cosmic Story*. That research opened a massive window into my understanding that there was much more to planetary habitability than what we see in our solar system. In realizing that our Sun, a humdrum G-Type star, isn't as unique as we've assumed, and that there are many other types of stars out there with potential to support life, I began to focus on those other star systems. An overview of the culmination all that research came together in this book.

The actual writing of a book is a fraction of the battle with editing making up the bulk of the work. Much like *Our Cosmic Story*, there were more drafts than words. My editor on that book taught me one of the most important lessons I'll remember, and that is to take the time to write a proper outline before jumping in and writing the story you are about to tell. This can be somewhat different than how you write a book around a fictional world or collection of ideas that demand inspiration first and facts second. Without a detailed outline to kick off the writing process, and a proper editing phase to complete the book, the facts and details you are trying to express may miss some of the most important points you wish to convey to the reader.

Any sort of writing is often seen as a solitary affair, but it has been anything but that in my experience. The more valued contributions have come from family, friends, my newsletter list, and acquaintances on social media. I am thankful for the patience of everyone, especially those who let me pester them to read a section or look over a phrase, often when the subject was merely an initial thought written down in haste. Thanks especially to: Ammy Sriyunyongwat, Cindy Anderson, Cora Nelson, David Zhong, Gary Nielson, Jessica Anderson, Richard Garriott, Scott Jennings, Scott Jones, Starr Long, Tin Khuong, Tony Medrano, and the scientists and other experts that gave their time for my often-lengthy questions.

Mathew C. Anderson

TABLE OF CONTENTS

INTRODUCTION 5
COMMON DEFINITIONS 9
HABITABLE EXOPLANETS: RED DWARF SYSTEMS LIKE TRAPPIST-1 11
INTRODUCING RED DWARF STARS 15
SURPRISE! RED DWARFS HOST PLANETS TOO 17
RED DWARF PLANETARY SYSTEMS 20
MAKING A RED DWARF PLANET HABITABLE 23
A STAR'S TEMPERAMENT 24
A PLANET'S SHIELDING 27
WATER WORLDS 36
A DELICATE BALANCE 40
INTRODUCING TIDALLY LOCKED WORLDS 43
CIVILIZATION ON A TIDALLY LOCKED WORLD 47
A NEW GENERATION OF TELESCOPES 49
SUMMARY: A UNIVERSE FILLED WITH HABITABLE EXOPLANETS 53
SUGGESTIONS FOR FURTHER READING 56
EMBEDDED REFERENCES 57
TERM REFERENCES 59
SCIENTIFIC PAPERS 61

INTRODUCTION

Finding exoplanets is far more challenging than finding the proverbial needle in a haystack. It has taken centuries of advances to achieve the technology and knowledge required to confirm worlds orbiting stars that are light years away. Only since 1992 have we had proof that our solar system isn't alone in the Universe. Now we know of hundreds of others and estimate that billions more exist in the Milky Way galaxy alone. The current picture being revealed through the latest state-of-the-art planet hunting telescopes shows us a Universe filled with the potential for life and civilization, even around stars we thought just a mere decade ago were inhospitable to life.

Our ability to confirm planets is heavily influenced by the size of the planet, its parent star, how far out it orbits from that star, and of course the distance of the entire system from Earth. The smaller and further out a planet is positioned, the more uncertain we are that it is in fact a planet and not some anomaly like a cloud of comets. With just a couple of star types that we can disregard in supporting more complex multicellular life forms like ourselves (these stars are Type-O, -B, and -A, due mainly to their short lifetimes of a billion years or less), over 90% of the rest are open to the possibility of supporting life, and over 70% of these in one type of star alone. Spoiler alert – the star type in question is not anything like our Sun!

In *Habitable Exoplanets: Red Dwarf Systems Like TRAPPIST-1*, you will learn about some of the constraints, as well as potential benefits, of habitable planets around the most abundant type of star in the Universe – red dwarfs (comprising of both late K-type and all M-type stars). Their abundance comes from the fact that these stars are comparatively tiny, and thus much easier for the Universe to produce. Small as they may be, they have some interesting attributes that both cast doubt and hope on their ability to host life. While life has been located around our own star type (as we are here to talk about that revelation), it would be shortsighted to ignore the vast majority of stars because some uncertainties about them remain.

To aid in our understanding of how alien red dwarf systems can be to our solar system, I will be frequently referring to TRAPPIST-1 and the discovery of seven planets in that system. TRAPPIST-1's configuration is optimal to allow scientists to test many of our assumptions on how planets form, and whether they can retain habitable conditions for the billions of years required for life to evolve. You will discover how a star both helps and hinders the evolution of life, and how any orbiting planets can remain in a state of environmental balance to allow life to grow and evolve. We will also explore one particularly unique quality of red dwarf worlds that may provide an even greater capability for supporting life than we have here on Earth.

A planet's distance to its star, as well as its size, density, and composition, are the primary considerations in determining whether it is setup to support other life-giving attributes like an atmosphere and surface liquid water. These features will also determine if the planet can form a protective magnetic field. For example, Earth is large enough for its interior to create a fluidic dynamo – the physical motion of material – in the outer core, which generates a globally encompassing magnetic field. This field keeps the solar wind at bay, preventing the oxygen-rich atmosphere from bleeding away into space. Even the tamest red dwarf stars are going to toss massive amounts of radiation towards their planets, many times greater than what Earth experiences with our Sun. A magnetic field is the first line of defense against this barrage.

An atmosphere plays many roles in a planet being hospitable to life. Without an atmosphere, all exposed liquid water would quickly evaporate into space. The mass and gravity of a planet will have a strong influence on atmospheric composition. Earth's atmosphere is composed of 78% nitrogen, 21% oxygen, and other trace elements. Oxygen levels in the atmosphere have fluctuated in the past, from as low as nearly 0% to as high as 35%. Oxygen also made – and maintains – the ozone layer, which is critical to shielding life from harmful UV radiation from the Sun. We don't yet understand what red dwarf planet atmospheres contain in detail, but ozone is thought to be one strong possibility for a couple of reasons we'll explore later.

Water is the most directly impactful attribute to consider when gauging habitability. On Earth, wherever water is found, life is also found (it's important to note that this observation is aided by the fact that all the water we sample shares a global biosphere). Life is thought to have first appeared in a mix of chemicals and water under intense heat, possibly deep within the oceans along volcanic ridges. Large bodies of water also help to regulate atmospheric temperatures and ensure a robust hydrologic cycle that seeds water onto otherwise dry land. There should not be too much water though, as this would result in a water world (a planet covered 100% by a single ocean). Water covers 71% of Earth's surface, providing enough land area for life to evolve into more complex forms, including being able to build a civilization capable of advancing technologically.

There are also exotic attributes of some exoplanets that call into question whether life is possible at all on these worlds. One particularly special attribute is what's known as a tidally locked position. If the planet is close enough to its parent star, it may be so strongly bound as to not rotate at all with respect to that star. This effect is exactly why our moon always shows the same face toward Earth as it orbits. If it's found that life is possible on such a planet, that life would experience a world unlike anything we experience on Earth. We will explore in detail this idea toward the end of the book, and what it means for the potential of life and civilization.

The remarkable achievement of detecting exoplanets is only the beginning of what will be realized in the decades to come. We live at a time where amateur astronomers can even get in on the action and discover entirely new attributes about these worlds. When a candidate planet is confirmed, we often see the contribution of an amateur astronomer being recognized. One such example is Andrew Grey's discovery of a system with at least four worlds in a size class called super-Earths (slightly

Mathew C. Anderson

larger than Earth). Mr. Grey accomplished his discovery not with multimillion-dollar equipment, but by sifting through endless streams of data stored in archives around the world. What computers and scientists may have missed is open for you as the ultimate treasure hunt of a lifetime. You can start the hunt (after reading this book of course) by joining planethunters.org to learn how others analyze the data, just like Mr. Grey has done.

There have been hundreds of papers over the last few years that have talked about exoplanets. Scientists have begun studying in-depth the idea of red dwarf planets having a protective magnetic field, for there to be an atmosphere with life-friendly gases and temperature ranges, and liquid water on their surface. While the knowledge we have gained is still highly evolving, it is quickly beginning to consolidate to provide us an intriguing picture of habitability. As soon as we can back up theories with additional evidence from the next generation of telescopes, we hope to have some definitive answers. This short and concise book will go into more detail on the above considerations and how they may apply to red dwarf worlds.

For a complete list of supporting material, read the Embedded References, Term References, and Scientific Papers sections at the end of the book. There are important follow-up technical links that explain why the papers come to their conclusions. I would also suggest you pick up my previous books in the *Our Cosmic Story* series to get a wider view on how might life and civilization evolve on other worlds, and the chance for that life to advance enough to one day explore space as we have.

COMMON DEFINITIONS

The following definitions and comments can be found to vary, so I chose ones that best fit the book's subjects. Definitions are either from, or based on, the Oxford English Dictionary (OED), NASA.gov, Space.com, or other official sources:

Abiogenesis: The original evolution of life or living organisms from inorganic or inanimate substances.

Civilization: The process by which a society or place reaches an advanced stage of social development and organization.

Cold Trap: A concept in planetary sciences that describes an area cold enough to freeze (trap) volatiles (like water).

Electric Field: A region around a charged particle or object within which a force would be exerted on other charged particles or objects.

Exoplanet: 'A planet outside of our solar system' In the book, 'exoplanet' is only mentioned when I specifically want to point out this fact. Otherwise 'planet' is used in conjunction with other references to suggest the planet is beyond our solar system.

Habitable Zone: The Habitable Zone is the distance from a star where liquid water can exist on a planet's surface.

M-Dwarf (Red Dwarf): 'A dwarf star, ranging in mass from one-tenth to one-half the mass of the Sun, whose relatively cool surface temperature makes it appear red-orange in color.' For the sake of clarity, we will be using 'red dwarf' for the book.

Ozone Layer: A layer in the Earth's stratosphere containing a high concentration of ozone which absorbs most of the ultraviolet radiation reaching the Earth from the Sun.

Magnetic Field: A region around a magnetic material or a moving electric charge within which the force of magnetism acts.

Planetary System: 'A star or multiple stars that has orbiting planets.' This is not the same as a 'star system', which is a star or multiple stars that is not referenced in the context of planets.

Snow Line: Boundary in a star system at a sufficiently cool distance from all nearby stars where simple molecules condense (such as water or water vapor turning into ice).

Super-Earth: A super-Earth is a planet that is larger than Earth, but not so large that it has the properties of a gas giant like a hydrogen and helium atmosphere.

Terminator: The dividing line between the light and dark sides of a planetary body.

Tidal Locking: 'Where an object's orbital period matches its rotational period.' An example is Earth's moon always facing the same way toward Earth.

Ultraviolet Radiation (UV Radiation): The ultraviolet part of the spectrum. Ultraviolet radiation spans wavelengths from about 10 nm to 400 nm, and is an important component of sunlight.

Volatiles: (of a substance) easily evaporated at normal temperatures, such as petroleum ether, hexane, and benzene. Water is also considered a volatile.

HABITABLE EXOPLANETS: RED DWARF SYSTEMS LIKE TRAPPIST-1

"To consider the Earth as the only populated world in infinite space is as absurd as to assert that in an entire field sown with millet, only one grain will grow."
- Metrodorus of Chios, 4th century BCE.

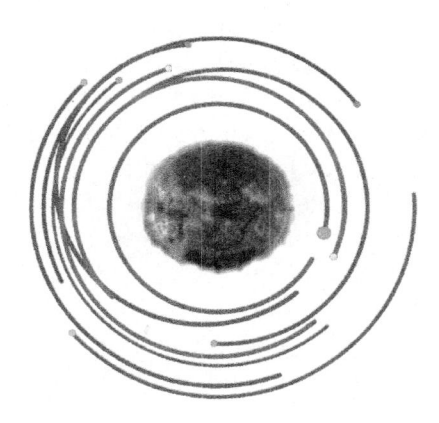

"Earth-like planet discovered just a few light years away!"
You may have seen sensational headlines like the above one with increasing frequency. Sophisticated telescopes have been gazing on distant star systems and have started to pick out the faint hint of orbiting planets. We are of course most interested in worlds like Earth, so the press tends to jump to the conclusion that we've discovered an actual other Earth with headlines like *"Earth-like planet discovered!"* or even *"An alien world right next door to Earth!"*, when scientist often only reveal the size and distance from its star are roughly similar. Nevertheless, incredibly exciting discoveries are being made that pave the way toward the day when we can truly declare, *"Earth-like planet discovered just a few light years away!"*

Mathew C. Anderson

The first exoplanet discovered was in 1992 by radio astronomers Aleksander Wolszczan and Dale Frail [1]. Two planets were in fact detected. While systems with multiple planets was not unexpected at the time, as our solar system has at least eight worlds, having them orbiting around a dead star known as a pulsar was especially surprising. Pulsars are often what remains of stars that have since exploded in a supernova. Before the discovery, astronomers expected planets would be destroyed within proximity to such an explosion. The current thinking is the system's original planets were indeed destroyed, with the two detected having been formed from the resulting debris. A true phoenix being reborn from its ashes.

A few years later in 1995, Michel Mayor and Didier Queloz of the University of Geneva announced the discovery of a planet around a G-type star, 51 Pegasi [2]. Scientists were ecstatic to have finally discovered a world orbiting a familiar type of star, and so soon after the first exoplanet in 1992. Before these worlds caught our attention, we weren't sure if *any* existed. Hundreds of billions of stars in our galaxy alone, and only one system and its single life-giving planet among them all? The proposition was unlikely in the extreme, but before these discoveries it was still possible. With 51 Pegasi, the search for life suddenly seemed much closer… at least until scientists got a better look at the position and composition of this seemingly *Earth-like* world.

Known as Pegasi b (the parent star is given the letter *a* with all orbiting planets labeled b, c, d, and so on), the planet was found orbiting its star just 7 million kilometers distant. Our solar system's Mercury orbits the Sun at about 46 million kilometers, and it still bakes in the Sun's thousands of kelvins temperature. This hellish furnace is hot enough to melt lead and instantly combust materials like wood or paper in the presence of oxygen. Pegasi b is so close to its star, and so hot, that it's puffed up like a balloon. If we could get an up-close look at the planet, it would glow a deep red. Further observations confirmed the planet was nowhere close to Earth-like and had little chance of harboring life – especially life that walks, talks, and can build a civilization.

Habitable Exoplanets: Red Dwarf Systems Like TRAPPIST-1

Our search for a truly habitable exoplanet did not start as we expected. The search up to this point was filled with a bounty of discoveries that quickly taught us the Universe is weirder than we ever imagined. The data suggests that there may be many more worlds of unexpected types than the ones we've discovered so far. As of 2018, scientists have confirmed the existence of thousands of exoplanets with hundreds of those being in the Earth and super-Earth categories [3]. Also, while planets the size of Mars and smaller are not found as often, it is highly likely that they are even more in abundance, we just haven't detected them yet due to their small size.

For years after 1995 we continued focusing on stars like our Sun – G-type (as well as close cousins of the slightly cooler K-type and slightly hotter F-type). The reason is understandable. We exist on a clearly habitable planet that's orbiting a certain type of star, and so life has at least some chance to exist around this type of star purely by direct observation. Statistics demand that other G-type stars then can, and will, at least in principle given a proper planet within the system, also host life. While these stars comprise less than 10% of all stars in the galaxy, there are still billions of them. This is more than enough to investigate to see if we are a common result of planet formation, or a strange statistical anomaly rarely (or never) found anywhere else.

What makes G-type stars so friendly to life? For starters, they last long enough before blowing up at the end of their existence (which will occur for the Sun in a safe and comfortable 4 billion years from now). Our Sun has been around for about 4.54 billion years with the Earth just a few million years shy of that number. This has given life plenty of time to evolve, and for that life to consider the chances of life elsewhere doing the same. Along with the Sun's longevity, it is a relatively stable star lacking frequent destructive flaring episodes. Flares can eject massive amounts of radiation and charged particles towards nearby planets, possibly destroying any life that attempts to make a foothold on those worlds.

The longevity of a star is also important for the ability of its planets to stay within the habitable zone (the distance around a star where liquid water can exist on the planet's surface). As the Sun ages and fuses increasingly more hydrogen into helium, its temperature rises. Most stars experience this process at varying rates. This is slowly causing Earth to be positioned more toward the inner edge of the habitable zone, eventually resulting in conditions on Earth closer to that of Venus in a far-off billion or so years. In the meantime, Earth's current distance to the Sun ensures that life will continue for at least another few hundred million years. Stability and longevity is key to allowing life the chance to evolve, so any factor that increases these conditions are going to be watched for with every exoplanet discovery.

Up until we started detecting exoplanets at an increasing rate, we disregarded the potential for life around the more exotic red dwarf stars, even if they are ubiquitous. With limited resources and the still-new field of exoplanet astronomy, it makes sense that we start with the type of star we know can host life. That is until technology can improve and research theories warrant the search for worlds around other types. These past few years has seen advances in technology and new research come together with a sharp eye toward red dwarf stars.

INTRODUCING RED DWARF STARS

The stars of most interest are those that are not newly born, nor in the process of dying. These stars are outside what is known as the 'main sequence', which is the period in a star's existence most stable and capable of supporting life (disregarding how long it resides in this phase, as you'll see below) [1]. The range of main sequence stars follow this order:

- O-Type: A very hot and gigantic star that shines blue and lasts for just 2-15 million years
- B-Type: A very hot star that also shines blue and lasts for about 15-250 million years
- A-Type: A very hot star that shines white and lasts for about 251-2,000 million years
- F-Type: A hot star that shines white-yellow and lasts for about 3-6 billion years
- G-Type: A star that shines yellow and lasts for about 7-15 billion years (like our Sun)
- K-Type: A slightly cooler star that shines orange to red and lasts for about 16-45 billion years
- M-Type: A cool star that lasts 46-700 billion or more years and numbers 70% of all stars in the Universe. They glow a deep red from a high output of infrared radiation.

As you can see, the term red dwarf is not listed above. The term includes both M-type and some K-type stars, referring to their red hued appearance, though also in how they use up their nuclear fuel. You may also see in literature the term 'M-dwarf', which are the M-type star specific red dwarfs. The terms are often interchanged, sometimes incorrectly, and even then, are admittedly a bit confusing. As long as you see 'M-type, 'M-dwarf', or 'red dwarf', then you are on secure ground for understanding how these stars behave and how they can support life-bearing planets. For the simplicity in what these stars mean for habitability, we will be keeping with the term 'red dwarf'.

Red dwarf stars are ubiquitous because the Universe tends to create more of the simpler or smaller of an object. For a cosmic example, the first atoms that formed after the Big Bang filled the Universe with 75% of the single proton element of hydrogen and 25% of the double proton element of helium. Red dwarfs are much smaller than our Sun by many times in radius. In fact, the smallest red dwarfs are about the same size as our planet Jupiter [2]. The reason that Jupiter itself hasn't become a star is because the planet's density is far too low to ignite a sustained fusion reaction. If Jupiter were between 75-85 times its current mass though, our solar system might very well have hosted two stars instead of the lonely one that shines in Earth's sky.

Red dwarfs are the longest-lived type of star possible, which is great news considering life needs a long period of time to evolve. The smallest red dwarfs will live for hundreds of billions of years, or well over 100 times the estimated lifetime of our Sun. To grasp the opportunity for life to have a trillion years to evolve, consider that a trillion years is about 73 times that of the Universe's age of 13.8 billion years. A trillion years is also 100 times the lifespan of our Sun, and 5 million times the total number of years humans have existed. You could watch Earth be born, witness the evolution of life, observe the sliver of time during which humanity builds great civilizations, eventually see the aging Sun boil away the planet's oceans and atmosphere, and then repeat the entire spectacle at least 185 times before the last red dwarf star finally burns itself out.

If, on average, it takes at least a billion years for a planet to become habitable, and another couple of billion more years for life to evolve into something capable of intelligence, red dwarf worlds have plenty of opportunity to develop the kind of life we are hoping to find. That is, of course, assuming there are habitable planets orbiting these stars.

SURPRISE! RED DWARFS HOST PLANETS TOO

The first red dwarf planet, Gliese 876 b, was discovered orbiting Gliese 876 in 1998 [1]. As of 2011, three more worlds were found orbiting the star. Unfortunately, the habitable zone planets (two of the four) are thought to be gas giants like Jupiter, and thus not habitable. These first red dwarf worlds may not be the last found there, and many other systems we've begun to detect. In fact, I wager that a high percentage of systems have at least one additional planet hiding in them. Those planets are either too far out in orbit or too small to detect with current technologies. At just 15 light years away though, this system is ripe for follow-up in the future as new telescopes come online.

To the delight of scientists around the world, we continue to discover red dwarf planets, and a lot more of them than we ever expected. As of 2018, hundreds of exoplanets have been found around multiple star types with the nearest system being right on our doorstep – Proxima Centauri [2]. The special property of Proxima Centauri is that it is a triple star system. For a long while it was thought that binary or trinary star systems would make it impossible for planets to form and remain stable in orbit around any of their stars. Over the last few years we've discovered that even these systems have planets. The Universe is continually revealing one surprise after another, forcing us to cross these surprises off our list of 'impossible'.

The vast majority of discoveries are thanks to the Kepler space telescope, including the first Earth-sized exoplanet, Kepler-186f [3]. While 500 light years away, Kepler-186f was detectable due to Kepler being specifically designed to find smaller worlds, and worlds in close-in orbits around their stars. Even more impressive was the fact that this system contained multiple Earth-sized worlds that included at least two super-Earths. Kepler-186f orbits its star within the habitable zone, though it's on the outer edge, like Mars is in our solar system. Does that mean it's a Mars look-a-like? Perhaps, though the planet's super-Earth status may well have saved it from a similar fate.

The most astounding exoplanet discovery to-date brings us to the TRAPPIST-1 system, and the system we will be frequently referring to in this book [4]. TRAPPIST-1 contains multiple worlds all packed together in tightly knit orbits around a red dwarf star. If you superimposed the system over our own, the star and its planets would fit well within the orbit of Mercury. These worlds are all similar in size to Earth, and at least three of them reside within the star's habitable zone. While this doesn't mean they are habitable (again referring to Mars), it does provide reason to investigate them further, and investigate we have.

The first planets around TRAPPIST-1 were discovered in 2015 with three orbiting worlds [5]. This was exciting enough to confirm around such a small star, but only a couple of years later was it announced that at least seven planets were in orbit. Wow! That's nearly as many planets as we have in our solar system. Only the Kepler-90 system has more, tying our own at a total of eight planets. What's most surprising though is that all seven of the TRAPPIST-1 planets are Earth-sized, ranging from a mass of about 33% to over 150% the mass of Earth. That's a remarkably narrow range for so many planets in one system. For example, our solar system has a mass range from a tiny 5% (Mercury) to a gigantic 318 times the mass of Earth (Jupiter).

After factoring for mass and density, the composition of a planet is key to its surface conditions, and initial data suggests the TRAPPIST-1 planets have a wide variety of compositions. The estimated densities alone suggest several of the planets are much lighter than what would be expected for a world like Earth, explainable with a thick atmosphere, or more likely covered in an abundance of water. This could mean the surfaces range from global ice fields to water worlds. There are many properties of these worlds that are ideal for use as a point of reference to describe some of the habitability processes we will later explore.

"No one ever would have expected to find a system like this. They've all experienced the same stellar history because they orbit the same star. It's a goldmine for the characterization of Earth-sized worlds." - Hannah Wakeford

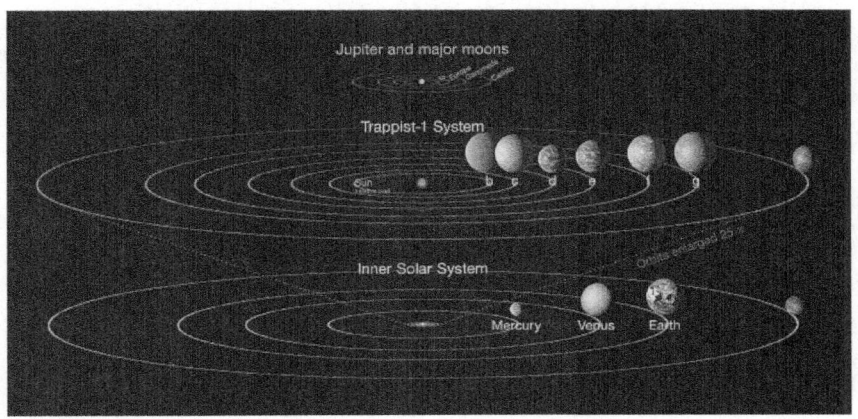

Habitable Exoplanets: Red Dwarf Systems Like TRAPPIST-1

(NASA/JPL-Caltech)

Planet	**Orbit**	**Radius**	**Mass**	**Density**
TRAPPIST-1b	1.51 d	1.12	1.02	0.73
TRAPPIST-1c	2.42 d	1.10	1.16	0.96
TRAPPIST-1d	4.05 d	0.78	0.30	0.48
TRAPPIST-1e	6.10 d	0.91	0.77	0.93
TRAPPIST-1f	9.21 d	1.05	0.93	0.85
TRAPPIST-1g	12.36 d	1.15	1.15	0.87
TRAPPIST-1h	18.76 d	0.77	0.33	0.55

Statistical estimates relative to Earth as of February 2018

Mathew C. Anderson

RED DWARF PLANETARY SYSTEMS

The current operating phase of the Kepler space telescope, called 'K2' and having begun in late 2013, is the phase in which most red dwarf worlds have been discovered [1]. Back in 2013 Kepler lost the ability to align with stars to track planets crossing the star field. This forced scientists to adjust the spacecraft to keep it operational, but the resolution they came up with resulted in another limitation. The new configuration only made it feasible to study stars that would likely have exoplanets orbiting much closer than Earth is to our Sun. Even so, the science that has been achieved since then is remarkable.

There are several common elements that many red dwarf systems are beginning to reveal to us. At first, we weren't sure if any planets at all existed around such small stars, but surprisingly most them have at least two planets, and possibly many more that are orbiting further out. Having more planets in a system naturally increases the likelihood that one of them will reside within the star's habitable zone. This also increases the chance that it will be Earth-sized as well. On this second point, these systems seem to prefer Earth or super-Earth sized worlds, just like what was revealed above for the TRAPPIST-1 system. The comparison is like how gas giants prefer moons of a relatively small size (at least in our solar system) compared to something like the size of the Earth.

Like all planetary systems, no matter the star type, planets form out of clumps of gas and dust. Eventually these clumps grow to such a degree that their gravity shapes them into a sphere for any object larger than 400-600km in diameter, or the size of the asteroid belt's Ceres body [2]. As the gas and dust forms into asteroids, comets, moons, and planets, they play a billiards game with each other, jostling for a stable position in the system. This will mean frequent collisions, as well as strong gravitational tugs that toss some of the newly formed planets out of the system entirely, or worse, toss them straight into the parent star. Eventually an equilibrium is reached between the remaining bodies. Some will remain farther out like the gas giants in our solar system, while others will migrate their way inward.

The hotter the star, the further out the planets will need to be, and G-type stars like our Sun will not support habitable planets much closer than what Earth is already. Because red dwarf stars are much cooler, their planets need to be close-in if there is any chance for life to evolve there. While planets in close-in orbits are not exclusive to red dwarfs, they do seem to be a common feature of these systems. Why doesn't our solar system have such close-in planets with a similar pattern of migration? It may have at one point in its history, but for reasons that scientists think relates to having gas giants in the outer regions, the migration either halted, tossed the remaining planets into the Sun, or ejected them out of the solar system entirely.

While havoc is destined to ensue at a certain distance between orbiting bodies, the threshold isn't as obvious as we might assume. There is a computer game called Universal Sandbox that allows you to simulate star systems and a range of other Universe properties, like the gravitational effects of a black hole, or what would happen if two stars collided [3]. In preparing the topics for this book, I decided to get the game and spend days (admittedly with a lot of fun destroying various objects) simulating several system configurations. One common result that occurred in nearly every simulation of tightly packed orbits were the planets' eventual demise. Even a system that seemed stable for millions of years would eventually become unraveled and planets would collide at some point. Sometimes this occurred in as little as a few millions of years after the system formed.

For long-lived and tightly packed systems like TRAPPIST-1 to remain stable, a specific configuration between the orbits of the planets must be achieved. It can only be possible to setup such a system if the planets initially formed further out and slowly migrated in over millions of years. As these planets migrated, their respective orbits nudged each other just enough to force a separation between them. This separation developed into a natural resonance that has persisted for billions of years. We are not yet sure if this is a common feature of red dwarf systems. It isn't required to fit a planet or two within the star's habitable zone, but it is certainly a helpful one to keep planets around long enough for life to evolve on them.

"If you were on the surface of one of these planets you would have a wonderful view of the other planets," Michaël Gillon, an astronomer at the University of Liège in Belgium and an author on the new paper, said in describing the discovery. *"You wouldn't see them like we see Venus or Mars, like dots of light. You would see them really as we see the moon. ... You would see the structures on these worlds."* – Michael Gillon, Astronomer at the University of Liege in Belgium

What we have learned in the past few years studying red dwarf systems is just how helpful this migration process might be for finding habitable planets beyond their orbital longevity [4]. The key to all life on Earth is liquid water. While we are not certain where the Earth's water inventory entirely came from, a portion is likely to have arrived as incoming ice-laden comets and asteroids. These bodies are particularly abundant in the outer reaches of the solar system, far from the Sun's ice melting heat. If an Earth-sized planet were to form at this distance around another star, its water inventory may end up being significant, even far greater than any body in our solar system. This migration process may also give the planet time to avoid the parent star's more temperamental early years, as we'll explore in the next section.

MAKING A RED DWARF PLANET HABITABLE

Now that we know that red dwarf planets are in abundance, far dominating the count of terrestrial-like worlds, let's dig in and understand the potential pitfalls of these systems hosting life. With stars so small and planets needing to be so close-in, what could stop life in its tracks before it gets a chance to sprout? The answers brought up next will give you both cause for alarm, excitement, and perhaps nudge your imagination into the realm of the wild and mystifying at how some of these worlds may even be more habitable than Earth.

There are three main categories of considerations we will go through to try and come to an understanding of habitability. The first is to take a close look at the parent star. As we know very well with the Sun, its stability is key to ensuring life isn't snuffed out at the first solar hiccup that occurs. The second consideration is focusing attention on any orbiting planets and how they may be able to protect themselves when the star does become restless. The final consideration is exploring the delicate balance of processes directly on, and within, these worlds in how they may regulate their atmospheres and keep their climates stable.

Mathew C. Anderson

A STAR'S TEMPERAMENT

All main sequence stars go through their early years as temperamental little brats. They spout off dangerous flares and coronal mass ejections that toss high-speed radiative material towards the outer system. This stellar material then rams right up against any orbiting planets and their delicate atmospheres. Smaller stars will generally see an exasperation of these more temperamental years, and red dwarfs are the worst of the lot. Their first 500 million years or so are bustling with activity, spouting off super powerful flares over a thousand times more energetic than what our Sun produces [1].

Flares occur when the plasma of a star heats up to millions of kelvins, driven by the star's magnetic field connecting to lower temperature spots on the surface. These spots are referred to as sunspots for our Sun. Since red dwarfs have fully convective interiors, that is their entire interior is a churning mass of plasma, their magnetic field lines can become extremely twisted. This twisting will channel huge amounts of energy to the star's surface. Much of that energy will shoot out toward nearby planets that may have little protection against the flare's effects. Daily bouts of super powerful flares are not uncommon for these stars, while our Sun may only spout off a relatively powerful flare every few months or even every few years.

Not all red dwarfs will continue to exhibit this extreme flaring after their early years, and even if they do, the flares are going to be somewhat different than for larger stars. While the flares may be more frequent and far more powerful than anything our Sun produces, the time length of the flares is much briefer, limiting the effect on an orbiting planet and allowing that world more time to recover before the next flare arrives. These flares are also suspected to be mainly proton-free, another limitation to their energetic impact [2]. Proton flares eject protons (and sometimes helium and other nucleic ions), which are highly energetic particles that accompany the barrage of radiation. These barrages are what give the Earth its spectacular auroral polar light shows.

Some of the radiation emitted by a star is in the form of Ultraviolet (UV) [3]. UV radiation is part of the electromagnetic spectrum between 400 and 10 nanometers with wavelengths longer than X-rays, but shorter than visible light. There are three types of UV radiation within this spectrum. UV-A (320-400nm), UV-B (290-320nm), and UV-C (100-290nm). UV-A is the longest wavelength of the three and the least harmful to life (though still quite dangerous, as it can pass directly through the atmosphere of a planet to the surface). UV-B would normally pass through Earth's atmosphere if it weren't for the protective ozone layer. At worst, UV-B typically causes a severe sunburn or facial wrinkle. UV-C on the other hand is extremely harmful to life, but fortunately is blocked by the atmosphere.

UV radiation played a big role in life's emergence on Earth, and almost certainly will on other worlds [4]. UV radiation is often viewed as harmful because it can disrupt molecules through photolysis, which is the process of molecules being broken apart and recombining. This can occur with RNA and DNA, causing mutations and rendering life forms unable to reproduce viable offspring, or worse, outright killing the organism. This process can also form toxic substances like hydrogen peroxide. Earth was thought to have life only in the oceans for the first few billion years because UV radiation cannot deeply penetrate water. Life thrived just a few meters below the surface before structures like the ozone layer came about to block off some of the incoming radiation, allowing life to finally stake a claim on the surface.

While UV radiation is generally shunned once life is formed and can reproduce on its own, this radiation was important when the very first life appeared. Known as abiogenesis, life may have depended upon a strong flux of UV radiation to split apart molecules that then recombined to form the first strands of RNA and DNA. Even lightning and other energetic sources are less effective to develop this prebiotic chemistry (though these other methods may still be sufficient on UV weak worlds). One tenet of the definition of life is the idea of constant change through cycles of formation, growth, decay, and eventually death. UV radiation provides a catalyst for this change to occur.

This is where red dwarf planets start to get interesting. We thought at first that UV radiation would be so strong that it would prevent even prebiotic chemistry from developing [5]. However, red dwarf stars emit a different mix of UV-A, UV-B, and UV-C radiation than our Sun. That balance tilts heavily in favor of there being less of the destructive form. *"At short wavelengths, M-dwarfs emit proportionately more radiation than the young Sun, but fluence at these wavelengths is robustly blocked by a range of atmospheric absorbers, including CO2 and H2O, which shield out < 200 nm radiation. Consequently, M-dwarf planets, so long as they can retain their atmospheres, are low-UV environments."* – Sukrit Ranjan, Robin Wordsworth, and Dimitar D. Sasselov from an August 1, 2017 paper [6]. However, this amount is so much less that it may prevent life from forming for the opposite reason of it being too destructive – it doesn't jostle molecules enough to build life's prebiotic needs. The one redeeming quality here of red dwarf stars is their infrequent proton flares may still produce the needed amount of UV radiation.

Once a sufficient quantity of prebiotic chemistry builds up, life can then find other niches to allow more complex life to evolve. If a red dwarf planet can survive the parent star's early temperamental years, that more complex life need only wait until the star settles down, and then it can further evolve into larger multicellular forms, eventually overtaking the land surfaces (if any land exists on the world). Life's appearance may be delayed on a red dwarf planet by a billion or more years than what occurred on Earth, but with up to hundreds of billions of years available, these worlds have all the time they need.

A PLANET'S SHIELDING

Now that we have an idea of how different red dwarf stars are compared to the Sun, what does an orbiting planet have available to defend life from such a hostile parent? Fortunately, there are several defenses that will aid in protecting life, especially any creatures swimming deep within the planet's oceans. We'll explore three key qualities, starting with a planetary magnetic field (or possibly the lack thereof).

A Protective Magnetic Field

"A magnetic field is a region around a magnetic material or a moving electric charge within which the force of magnetism acts." (OED). In the case of planets, a magnetic field is generated in two ways. The intrinsic way is in the interior of the body. The field arises due to the fluidic nature of the inner material as it convects heat outward toward the surface. The other type of magnetic field is one generated through the interaction of UV radiation coming from the star hitting the planet's upper atmosphere. This interaction creates ionizing radiation and an ionosphere which acts to push back the stellar wind, creating an externally induced field and shielding the atmosphere from further degradation. Both field types work to stop the star's charged particles from impacting the atmosphere and surface.

The rotation rate of a body also plays an important role in generating an intrinsic field by centrical force upon the body's interior. The faster a planet spins will increase the chance its field will be stronger. Planets in the habitable zone around red dwarf stars are all suspected to rotate slowly, or even synchronously with their star. We'll get more into this intriguing synchronicity phenomenon a bit later, but for now the point is that these planets may not rotate fast enough to generate a strong enough intrinsic field. They could still have one in a weaker form, or an externally induced one, or possibly lack both options entirely. The latter case would mean almost certain loss of the planet's atmosphere and water as the stellar wind strips it away.

Regardless of the strength or type of a magnetic field, they appear to be common around planets, as evidenced by planet's in our solar system. While Mercury lacks a thick atmosphere, it still has a weak intrinsic field due to its slow rotation and sluggish interior. Venus's magnetic field is externally induced due to its thick atmosphere and lack of an internal dynamo. Mars appears to have both a weak externally induced field, as well as additional weak localized 'mini-magnetospheres' scattered across the surface. The outer planets have fields at least the strength of Earth's, and in the case of Jupiter, the strongest of them all. Considering this diversity in our solar system, magnetic fields are suspected around exoplanets as well.

Some papers suggest close-in planets like those in the TRAPPIST-1 system may have magnetic fields that connect with the star's own field [1]. This would effectively leave the planet unprotected from the star's stellar wind as it is channeled along the magnetic field lines straight toward the planet's poles. While this suggests a blow to finding habitable worlds, most of this research points to just the inner most worlds being affected. For the TRAPPIST-1 system, only the inner two worlds may have this effect, and they are not even in the habitable zone anyway. This distinction is important when you read articles or papers written about exoplanets. An article will suggest these worlds are uninhabitable, and the next will suggest they are habitable. The key is to not focus on a specific scenario or property, but take them all into account in one big picture view. That's what we're doing here in this book.

The latest research clearly suggests that a magnetic field of some kind is a requirement for planets to be habitable [2]. Another paper emphasizes that even weak externally induced fields are still strong enough to protect the atmosphere and block certain types of radiation [3]. And the intrinsic field type may also be strengthened due to gravitational effects of the star constantly tugging on the planet as it orbits, especially for these close-in worlds we've been talking about.

A Protective Atmosphere

Assuming a planet has a magnetic field of some kind, its atmosphere is the next consideration in supporting life. This is primarily because an atmosphere can have properties that protect any surface water that would otherwise evaporate away into space (with some important exceptions like what happened with Venus). An atmosphere also acts as a thermal blanket, helping to maintain a consistent temperature across the planet. Weather processes, oxygen for life's energy needs, and many other components of an atmosphere are critical for the sustaining of life. The atmosphere can also be altered by life's very own processes (through both natural and artificial means), as we witness with Earth's biosphere.

Gravity also plays an important role in both retaining an atmosphere and the composition of that atmosphere. Hydrogen and helium are the lightest elements in the Universe, so only planets 1.6x the radius or larger than Earth will likely retain these elements in their atmospheres [4]. Fortunately, we don't need them to breathe, and indeed Earth's atmosphere contains just trace amounts. However, if the planet is too small, even with a strong magnetic field, all elementary gases (including oxygen) will still leak into space due to a weak gravity field. The minimum mass of a planet to sustain features like an intrinsic magnetic field and a habitable atmosphere in a star's habitable zone is about 1/3 the mass of the Earth [5]. Mars has $1/10^{th}$ the mass of the Earth, and thus is the leading reason why the planet is in such a desiccated state today.

Examples of the potential diversity of atmospheres around exoplanets is provided right here in our solar system. Venus is a hellish version of Earth with an extremely dense atmosphere that contains almost entirely carbon dioxide (CO_2). A bit further from Earth, we have poor little Mars that couldn't hold onto its atmosphere and is now as dry as the Atacama Desert and colder in the Martian winter than that of Antarctica. And the gas giants are just that – huge balls of hydrogen and helium gas compressed at various levels into other forms like metallic hydrogen. While these giants have an atmosphere, life would only be possible in the upper most layers, and that life would still need all the catalysts Earth is able to provide.

Atmospheres have been found on planets orbiting both G-type and red dwarf stars. They have also been found clinging to worlds that are as far away as the Earth is to the Sun, and so close to their star that the atmosphere of the planet literally exchanges material with the chromosphere of the star. Most of the examples have been outside the range of expected conditions of habitability, but that's a good sign. If a delicate atmosphere can exist in extreme environments, like what's suspected around the radiation soaked worlds of the TRAPPIST-1 system, then we may expect to find many more such worlds once new telescopes with improved detectors come online in the next few decades (more about these telescopes toward the end of the book).

While the parent star is a constant threat to a world's tenuous atmosphere, most atmospheric loss is estimated to occur well before the planet enters the habitable zone. The first and most significant loss will be the lightest gaseous elements of hydrogen and helium over the first few million years during the planet's formation. The rest of the bulk loss will happen well before completing a migration toward the inner system over the succeeding millions of years. If the planet has any atmosphere left at this point and hasn't been depleted further due to the lack of a magnetic field, or the planet's size is just too small, then atmospheric density is likely to be at least as dense as Earth's for a planet of similar size.

Assuming an atmosphere like what we have here on Earth, then we mainly need to understand how it protects life from the harshness of space. We already talked about how the atmosphere can absorb some of the incoming UV radiation, leaving perhaps just enough to pass through to aid, but not destroy life on the surface. The ozone layer was mentioned earlier as well, and is another key blocker of radiation that would harm life. The ozone layer in the Earth's atmosphere resides between the troposphere and the stratosphere several kilometers up. The layer exists in part because oxygen here doesn't like to stay by itself for very long. Oxygen molecules break apart due to radiation hitting them (photolyzed), and then recombine when encountering another oxygen molecule to form ozone (O_3) [6].

Oxygen wasn't always part of Earth's atmosphere. It took a couple of billion years for the gas to accumulate, continually generated mainly as a byproduct of single cellular life. After enough oxygen saturated the oceans and then started building up in the atmosphere, it was only at this point that the ozone layer formed through the above process. The layer then provided life a haven for the opportunity to evolve to new and more complex forms that could survive and thrive on the planet's surface. Like many processes on Earth, life evolved to indirectly protect future versions of itself, using the oceans and atmosphere as key ways to automatically regulate the necessary processes.

Life's involvement in forming an ozone layer isn't the only way ozone can be created though. Given a thick enough (but not overly thick) atmosphere dominated by CO_2, it's been suggested that oxygen could be produced in a sufficient quantity to create a naturally occurring layer [7]. The process starts with CO_2 being broken down by UV radiation, just like with the oxygen molecules being split apart. What remains is free carbon and oxygen. The carbon sticks around for a while as it doesn't like to react with other elements as much, but the oxygen quickly combines to form ozone. Since red dwarf stars emit greater levels of UV radiation than the Sun, the byproduct of oxygen would only be enhanced. This level of ozone could be many times denser than Earth's, protecting life on a red dwarf world.

As tenuous as we suspect atmospheres to be on red dwarf planets, the ability for these worlds to have an atmosphere at all gets even more exciting when we consider how a dense atmosphere may be beneficial to its own survival. The first consideration being that there's simply more gas the planet can afford to lose, stalling any eventual demise of surface water.

While volcanism and other processes may work to thicken an atmosphere, leakage of at least the lighter unwanted gases (hydrogen and helium) into space over time is inevitable. A dense atmosphere also allows heat in the upper layers to spread out, further regulating temperature extremes from the poles to the tropics. This is particularly important to prevent what's known as a snowball state, which we'll cover in more depth shortly.

In the case of the TRAPPIST-1 system, all the planets are near Earth-size and could have atmospheres similar in diversity to our inner solar system planets. They could be CO2-rich like Venus, have just the right balance of gases like Earth, or still have lost their atmospheres like what Mars had occur. Scientists are not yet sure if any of them have an atmosphere at all, but the measurements taken so far strongly infer several of the planets to have low densities. This points in favor of ocean worlds, and/or thick atmospheres filled with lighter gases like hydrogen and helium. The former is much more likely, given the smaller size of the planets. Even TRAPPIST-1d, a world between the size of Earth and Mars, may still be large enough to have sufficient gravity to hold onto its nitrogen and oxygen.

If an atmosphere is indeed possible on red dwarf planets, let's then look at how that atmosphere protects the most important asset of all to life - water.

The Blood of All Life - Water

Water is the blood of all life. For life to have evolved at all, a constrained but changing medium is needed. Solid materials are mostly unchanging, and conversely, gases are too freeform for cells and more complex molecules to collect and self-replicate. Liquids provide the best medium. Water's neutral PH and dissolution properties make it perfect for life. The oceans command Earth and team with organisms of all kinds, as do the rainforests and other biodiverse areas. The trait they share is a constrained but continually changing medium of water, minerals, and other elements.

If Earth's magnetic field and atmosphere were to disappear today, the oceans would instantly start to boil. Water on the surface would either completely evaporate away into space forever, or at some point the process would cool the water immediately below the surface, freezing it over and capping off further losses. This latter scenario could allow ocean life to continue to survive at the deepest points, but it depends on how the atmosphere is lost to begin with. Either way, a diversity of life and civilization anywhere near the likes of ours would effectively be impossible at this stage of a planet's demise.

There is both good and bad news regarding a planet keeping its water inventory long-term. First, the bad news. The parent star's UV strength still has an influence on water in the planet's atmosphere at certain heights. Water is a molecule that is made up of two hydrogen atoms and one oxygen atom. Evaporation plays a key role in how water molecules travel from oceans, lakes, and rivers, to traveling through the atmosphere to form clouds and eventually coming back down in the form of rain or snow. Water molecules can drift up to high levels where they are pulled apart by the star's UV radiation [8]. While water molecules themselves are too heavy to be pulled up into space, if they are broken apart, this allows especially the light hydrogen to escape.

The process of lifting these molecules up into the atmosphere can be assisted by a planet's electric field [9]. Somewhat different than a magnetic field, the electric field is a differential in charge from one location to another. In the case of worlds like Earth, this charge differential goes from the ground upward to the ionosphere. Since atoms of a molecule have a charge of their own, this allows them to be uplifted by the planet's electric field, providing the opportunity for them to be carried right out of the atmosphere by the stellar wind. Once an atom leaves the planet's gravity well, it is effectively lost forever.

Venus has an especially strong electric field. While scientists are not sure why the planet has such a strong field, it's suspected to be related to the planet being so close to the Sun. With a strong enough electric field, even the heavier oxygen atoms will be carried away into space. This process slowly seeped away

Venus's water over eons, leaving it a desiccated world today. TRAPPIST-1b and TRAPPIST-1c receive similar levels of UV radiation, so their electric fields may also be similarly high. They are already regarded as being outside the habitable zone though. The other TRAPPIST-1 worlds may be sufficiently far from their star to avoid developing a high electric field, but further observations will be needed to confirm this hypothesis.

There is good news for a planet with a strong electric field to be able to retain its water inventory. As you reach higher atmospheric levels, temperatures will fall and rise at certain points. Suspected to be the case on many worlds, and confirmed on Earth, there is a lower layer called the troposphere where temperatures plummet by tens of degrees. As the temperature drops, ascending gases are forced to drop back to Earth due to their high melting points. This layer is called a 'cold trap' [10]. Water forms ice crystals at the cold trap boundary, which then succumb to gravity, dropping back and eventually reforming as rain droplets or snow. This process is just below the higher UV radiation threshold and is thus very efficient in preventing water from escaping the planet, even with an electric field that lifts molecules to higher elevations.

Venus and Mars do not have cold traps. Mars lost its water because its atmosphere was lost due to the planet's small size, and it has a different crustal composition that is thought to have soaked up the remaining water. Venus on the other hand still has a thick atmosphere, and yet is more bone dry than nearly every other world in the solar system. The reason for its lack of a cold trap stems directly from the planet being too close to the Sun. Even if its electric field was non-existent, the lack of a cold trap would be enough for it to lose its water inventory. Equivalent examples to Venus and Mars here continue to be TRAPPIST-1b, likely TRAPPIST-1c, and possibly TRAPPIST-1d, depending upon how the planets initially formed.

Let's look at how Earth's atmosphere is structured. Within the upper layers you have the stratosphere. This layer is where the electric field and interactions with the solar wind are going to be the strongest. Any water molecules that reach up to this point are almost certainly lost to space after being split apart into hydrogen and oxygen. The first line of defense against this loss is the ozone layer toward the bottom of the stratosphere and top of the troposphere. Further down we have the cold trap in the upper troposphere just below the ozone layer. Even a strong electric field here won't matter much because the water molecules won't get a chance to get high enough to disassociate. They will instead drop back down to the surface. If the order and strength of these layers is similar on exoplanets, UV soaked worlds around red dwarf stars like TRAPPIST-1 may be comfortably protected from losing their water inventories for billions of years.

Mathew C. Anderson

WATER WORLDS

It's been theorized that Earth-sized and super-Earth sized worlds could retain many oceans worth of water that completely cover the planet's surface [1]. The further a planet is from its star the colder it gets, and thus more volatiles like water can remain bound to it. Even if much of the water was lost in the migration process toward the inner system, the planet could still end up being a water world. That global ocean could even be hundreds of kilometers in depth. By comparison, Earth's oceans reach a maximum depth of 11 kilometers, which is still an extreme depth, at least for any sort of vessel. For example, modern nuclear attack submarines have a tested depth of about 487 meters, or just under half of one kilometer.

Several of the planets in the TRAPPIST-1 system have estimated densities considerably lower than what would be expected for an Earth-sized world of similar composition [2]. TRAPPIST-1f has a radius of just .4% greater than Earth's, yet the planet's density is nearly 33% less. TRAPPIST-1d and e also have lighter densities. The most likely solution is they are entirely covered in a global ocean – a water world. Whether that ocean is in liquid form or locked up in ice is another uncertainty. Liquid water weighs 1 gram per cubic centimeter (g/cm3). Earth's crust is between 2.7 to 3.0 g/cm3, the mantle is between 3.3 and 5.7 g/cm3, and the core is at least 12 g/cm3. You can see how the volume of a sphere can easily be filled with the lesser dense water to result in an overall lower density planet.

Scientists have strong evidence that suggests ocean worlds exist right here in our solar system – Europa, a moon orbiting Jupiter, and Enceladus, a moon orbiting Saturn, are suspected to contain vast oceans underneath their icy surfaces [3]. These are all worlds far beyond the orbit of Mars, past the boundary known as the 'snow line' where it is too cold to retain liquid water on the surface of a body. They are also all moons of a planet, which presents an intriguing alternative scenario for life outside the typical habitable zone. If life is discovered beneath these ice capped moons, we may have to revise our estimates on the extent of life (at least for simpler forms) elsewhere in the Universe.

Believe it or not, Earth is a water-poor world [4]. While 70% of the planet's surface is covered in this life-giving substance, what lies deep within its interior is what makes up the difference. When we consider the water inventory both on the surface and within the mantle by the total mass of the planet itself, the percentage of water ends up being only about .1% of the total mass. If all this water were formed into a sphere, that sphere's diameter would be less than the state of Alaska's diameter. Even Europa's suspected water inventory is 2-3x the Earth's oceans. Earth is thus scientifically designated a 'dry' world, though clearly not a desiccated one.

A water world is even more likely than a planet that strikes a balance of water and land like we have on Earth. Large objects in the Universe are remarkably smooth when comparing their size to the surface topography, far smoother than any classroom globe's artificial topography would suggest. Earth is just about as flat as a bowling ball's surface when matched in size. That means that there wouldn't need to be very much more water than there is now to completely cover the land. It would only take a fraction of a percent of the total planet's mass to do the job. If you have ever spilled a gallon of milk across the floor, you know how far it can spread. This is a key reason why the Earth's melting glaciers and ice caps are such a threat to coastal cities.

Living in a global ocean sounds fascinating, but there are many consequences to life appearing and eventually being able to form a civilization there. Life depends on many more elements and molecules than simply water. Our DNA includes compounds called phosphates made of the element phosphorus, a highly reactive substance. Due to its reactivity (often with oxygen), it's rarely found as a free element, instead locked up in a variety of minerals like lithiophosphate, phosphammite, and the tooth mineral apatite. The only way life could naturally make use of this phosphorus is to break apart its surrounding minerals, which are in turn locked up in rocks. On Earth, the key process of unlocking the precious phosphorus is through silicate weathering (the process of CO2 being converted into rocks, and vice versa) [5]. UV light combined with rain and other weathering processes will wear down the exposed rocks and release the phosphates. A more limited process occurs at the bottom of the planet's oceans, and meteorites provide these minerals as well.

A global ocean would severely limit the process of releasing life giving elements, and it gets worse the deeper the ocean. With a planet containing just 1% of Earth's mass in water (or about fifty oceans worth), high pressure ices would form kilometers down, locking in the silicate rocks on the ocean bed. With depths of 2% or greater, the tectonic processes that provide other nutrient rich benefits via volcanism may stagnate or never get started [6]. In short, a planet with too much water could be as dead as a planet without any water whatsoever. We need to look for worlds that have both the right distance from their star to keep water in liquid form, but also be the right density to allow just a small fraction of water on the surface.

Let's say though that an ocean world has an abundance of free phosphate compounds and life abounds there. This best-case scenario still has a problem when that life becomes intelligent and decides to reach for the stars. There are countless steps of knowledge and technological progress that must be attained before reaching a level of civilization the likes of our own. For example, there is no way to develop a radio telescope without the knowledge and materials to conduct electricity, process signals, and house the equipment. The production and maintenance of such equipment will also need thousands or even millions of individuals through the supply chain. Pick any area of our technological civilization and see if you can trace back the steps to realize how we got to where we are today.

With a global ocean that is dozens of kilometers in depth, all the resources and energy catalysts you would need to build even a rudimentary civilization would be too difficult to obtain, out of reach entirely, or even non-existent. Without these resources that Earth has in abundance, a technological civilization would be impossible to get started, or simply too difficult and lengthy for its citizens to ever justify the effort. Any intelligent species on such a water world would forever be a prisoner with no hope of realizing what was truly beyond their world's atmosphere, or beneath them in the great depths of their single global ocean.

A DELICATE BALANCE

For life to be maintained, new energy and material must be inputted, transformed as much as possible to be useful to the organism, and waste byproducts then outputted. The same can be said for how a planet maintains conditions suitable to life. With an atmosphere and water inventory that would need to survive for billions of years, keeping conditions stable enough is a complex process. In conjunction with life itself, the planet must have some way to add and remove gases as they become out of balance.

For example, as heat trapping CO_2 builds up in the atmosphere, the silicate weathering process increases due to the increased rainfall from the higher temperatures. Rainfall wears down exposed surface rocks and dumps the minerals into the oceans where ocean life then uses those minerals to turn the dissolved carbon into carbonate rock on the seafloor. CO_2 is also vital to plant life, just like oxygen is to humans. As CO_2 is sequestered through these processes, atmospheric levels drop, resulting in a lowering of the planet's temperature back to nominal levels. Rainfall and plant growth levels also continue to drop, until they reach such a low level as to cause CO_2 to once again build back up in the atmosphere.

Plate movements are also important to ensuring CO_2 has somewhere to go when too little or too much exists in the atmosphere [1]. While the resulting earthquakes of huge plates moving across Earth's surface (albeit at just centimeters per year) is not typically associated with something positive, the process has tremendous implications for life. As plates get subducted under other plates, that CO_2 carrying material melts and gets buried deep into the planet's mantle. In other areas, fresh material is erupted, releasing CO_2 into the atmosphere. Think of plate tectonics and volcanism as the heart and veins of your body, and the silicate weathering system as the pressure valves.

There are a couple of processes that can cause a planet to be pushed out of equilibrium and beyond the ability for the above systems to self-correct. The first is known as a moist greenhouse state. This is when the evaporation of a world's water is so extensive, the global atmosphere becomes something akin to a wet afternoon in the Amazon rainforest [2]. While planets further from their star will welcome more CO2 to keep warm, they may still succumb to a moist greenhouse state if they have an extreme excess of CO2. As was mentioned in the beginning of the book, red dwarf stars emit most of their radiation in the infrared. Like the coils in your oven heating a delicious dinner, infrared radiation is also excellent at heating a planet's atmosphere. Expectations of habitability will need to consider this variable.

Let's say a world is dominated by a global ocean with no landmasses, is in a warm greenhouse state, has a high level of atmospheric CO2, and has a weak cold trap due to the higher temperatures. As water is disassociated and lost into space, the surface water continually lowers until eventually land begins to poke up out of the ocean. That land will then erode as rain falls. As rocks become exposed, they soak up CO2 through the silicate weathering process described above. With even a modest 10% land area compared to Earth's surface, this may be enough to remove much of the CO2 from the atmosphere, cooling the planet and possibly halting the moist greenhouse state [3]. As the cooling continues, the now drier air causes an increase in desertification across the land areas. Since silicate weathering relies upon rainfall, the weathering effect described above lessens, slowing down the CO2 absorption process. A balance has been struck, one that may persist for billions of years with life's assistance, just like it has with Earth.

For Earth at its current position within the Sun's habitable zone, a Venus scenario would require atmospheric CO2 to be 10,000 ppm (parts per million) [4]. Carbon dioxide levels are currently at just over 400 ppm. As far as we can discern, Earth has never gone above 4,000 ppm in all its post-Hadean era history (its initial formation period). Fortunately for the future of life, even civilization's destructive nature will not be enough to raise carbon levels directly by more than a fraction of historical highs. There simply aren't enough sources like coal and oil to do the job. Naturally though, carbon levels can rise further with the melting of permafrost and volcanic outgassing. In any case, present life (humans included) will find it very difficult to adapt to an even modest long-term warming trend, if it can at all.

The corollary to the moist greenhouse is a global snowball state [5]. While a higher concentration of CO_2 would allow a planet to warm to have liquid water on its surface, there is a limit here. Even with an atmosphere like Venus's, dominated by CO_2, at some distance from the star the planet will simply be too distant to keep warm and will succumb to a snowball state like the Saturn moon of Enceladus. The entire world could be covered in a single multi-kilometer thick ice sheet. Earth has been in multiple snowball states during times of low CO_2 levels in the atmosphere and a high albedo (reflectivity of the surface that prevents heat from being absorbed), but has broken out due to volcanic outgassing and other effects. Earth's atmosphere is also somewhat thin when we look at a range of scenarios that might prevent a snowball state from occurring again. Since the last episode, our planet has managed to keep in overall balance, once again thanks to life's assistance.

Having extremes is not always bad though, if those extremes keep to within the bounds of habitability. The more important consideration is that change does in fact occur, allowing life to find a niche when the conditions are right, and then helping with the regulation process once that life has a chance to blossom. Without environmental pressures of some form, life of whatever state it is in may not feel pressured to change and advance in form. A peek at our history clearly suggests that there is a delicate balance between stability and change. The two snowball Earth scenarios identified in Earth's past caused massive changes in the biosphere. One such state allowed the Cambrian Explosion, and an earlier state allowed the buildup of oxygen in the atmosphere [6]. These two states were triggered from changes in nutrients in the oceans from the melting of ice, allowing life to blossom. Higher intelligence may not have evolved as well if there had been no need to migrate to other areas, which forced social interactions with other groups of early humans.

INTRODUCING TIDALLY LOCKED WORLDS

The most intriguing possibility of habitability has yet to be explored in this book until now. I've saved it for last due to its fantastical nature. To begin to understand how different the type of planet I am about to describe is from our experiences on Earth, let's first reflect on what a typical day is like for us. We may envision the Sun rising in the east, passing overhead for several hours, perhaps peaking behind a partly cloudy sky, and then settling beyond the horizon to the west. The day becomes night which lasts for another handful of hours. During this cycle, we will awake in the morning, go to work or school, and then come home in the evening. Day after day the process repeats as we remain relatively fixed to a local area on the planet.

All planets in the habitable zone around Sun-like or larger stars are going to experience the process of day and night cycling. This is due to every planet being in constant motion as it orbits its star from a distance, rotating like a spinning top. The rate of that spin will be determined by the initial characteristics of how the planet formed, and how it evolved over time. Some planets like Jupiter and Saturn will have days that last only half of what we experience on Earth. On the other hand, there are others like Venus that rotate retroactively, completing a day that's technically longer than its year. Life on any of these worlds will experience the day and night cycle, just at a different rate.

If you look up at our moon on a clear night and do this over the course of several evenings, you will notice that the same side faces Earth each evening. How could this be if everything is rotating? Shouldn't the moon be rotating as well? Yes, and it does, but we don't notice it because the motion is in sync with our planet's own rotation. Like two ballerinas staring into each other's eyes, they are barely aware that it's not the room that's spinning, but they themselves. Their motion is in what is called a tidally locked position [1]. This synchronicity results in the moon permanently facing the same way toward Earth.

Mathew C. Anderson

Whether it's a moon orbiting a planet, a planet orbiting a star, or even a star orbiting another star, they all will have a certain threshold where the tidal locking effect comes into play. The closer a planet is positioned to its star, the more likely it will succumb to the tidal locking effect. While Mercury is the only planet in our solar system that is tidally locked with the Sun, this may be the exception more than the rule for most other stars. In fact, it is beginning to look like the vast majority of systems, most of them featuring a red dwarf star, have tidally locked planets. Where Mercury orbits in 88 days around the Sun, red dwarf planets will orbit in a matter of a weeks or even hours.

Because a tidally locked planet forever faces its star the same way, no matter where an observer is positioned on the planet's surface, the day and night cycle we experience would not exist. There would be a point on the planet in direct alignment to the parent star, bathed in perpetual sunlight with the opposite side shrouded in perpetual night. One side would forever enjoy a sky covered in a deep red glow, and the other side a sky covered in an equally beautiful dark canvas of twinkling pinpoints of light. Little would change for a stationary observer, aside from the daily weather patterns of clouds and rain passing overhead.

When red dwarf planets were first confirmed, scientists thought that such worlds would bake on one side and lock up all their water into ice on the other side. Since water is a volatile, it's easily heated and able to be carried by the atmosphere to cooler areas where it can condense and freeze. Since these planets do not rotate with respect to their star, ice accumulation on the night side can be permanent. In as little as a few million years, this process could make a world impossible to foster life with all its water locked up in ice or lost to space. This is definitely not an ideal situation. Fortunately, simulations reveal several feedback factors that may keep some of these worlds from becoming half-snow cone/half-burnt toast [2].

The most important factor for a stable climate is again a planet's atmosphere. Convection processes through atmospheric winds and deep ocean currents can help to even out the global temperature. A thicker atmosphere, especially one with more greenhouse gases than we have on Earth, would aid in stabilizing the climate (but only on a world that is sufficiently distant from its parent star to avoid a moist greenhouse effect from taking over) [3]. Atmospheric pressure is measured in bars. For example, Earth is at 1 bar, whereas Venus is at 93 bars, or the same as being 1 kilometer underwater. A dense enough atmosphere could prevent the locking up of water, and the atmosphere itself, on the night side. Barring topographical features like mountain ranges and the effect of localized bodies of water, a tidally locked world may have very stable and predictable weather patterns.

Plants may even show narrow and elongated growth patterns as their seeds are tossed in consistent directions. On the other hand, this may also inhibit normal reproduction efforts with only a few directions to spread one's offspring, so other methods may result. Plants will also need to evolve ways of capturing more energy in a low-light environment, and lit from a star that emits more energy in the infrared. Our Sun produces plenty of energy that is useful for photosynthesis, so plants don't need to absorb every wavelength to get the energy they need. If they did though, they might appear black in color [4]. For example, most plants don't absorb the visible green part of the spectrum, instead reflecting it, which is why leaves are generally green in color. If you ever wore a black shirt on a hot summer day, you have an idea of how much of a thermal difference color can make!

One advantage for life on a tidally locked world is the night side would receive little UV radiation, since it is forever facing away from the parent star. If the star's flares do not end up stripping away the atmosphere, and abiogenesis is still able to occur, then life may flourish. Life on the night side would need to evolve ways of using energy from sources other than light, but we have plenty of examples of this process occurring right here on Earth. Chemicals in the water and land, active heat sources, and other life itself can provide all the sustenance needed. Then, at some point in the planet's future, that life may migrate to the day side land areas after evolving ways of protecting itself from the radiation. This is good news for planets that receive a larger amount of radiation and have a weak or non-existent ozone layer.

Earlier I mentioned the need of a planet to have either an intrinsic or externally induced magnetic field. The former may require a freely rotating body, and red dwarf planets are locked in rotation with their star. These worlds have a bonus feature though that may help. They can be close enough to their star for the tug of that star to stretch the planet ever so slightly as it orbits, like what occurs for the moons of Jupiter and Saturn. For Jupiter's moon Io, active volcanism covers the moon's surface due to this friction process. Known as tidal heating, it may be sufficient to generate a strong intrinsic magnetic field [5]. Especially in the case of more eccentric orbits (non-circular), it can also kick-start and strengthen plate tectonic activity and volcanism on worlds normally too small to have these processes occur on their own.

If a red dwarf planet that's billions of years old is found to have an atmosphere and contain liquid water on its surface, it has escaped its star's deadly flares. Confirming such a world will indicate that it has a good chance of supporting life, possibly for hundreds of billions of years. With the planet's orientation to its star being the same all year, seasons and even long-term ice ages will be infrequent or mild. Perhaps the most serious instability would be from within. Plate tectonics with volcanism and super volcanoes will likely still occur, causing temporary climate change for decades or more after the event. This occasional bout of instability though may be just what life needs to evolve.

CIVILIZATION ON A TIDALLY LOCKED WORLD

Of all the experiences that we have on Earth, there is nothing like how civilization would be forced to adapt on a world that's tidally locked. While it's been only a few years that we've had a hint of the conditions on red dwarf planets, every year has more research published that, for the most part, expands the habitability thresholds. The imagination stretches to its fullest in coming up with wild and fantastical civilizations that could exist on these worlds. Just imagine what it must be like with a day that never sets and a night that never sees the light of day. Imagine explorers like Marco Polo, Charles Darwin, and Ferdinand Magellan setting out on an ocean covered in perpetual light or darkness. As they continue their journey for weeks or months in the light, the light above slowly shifts as an eerie darkness across the horizon grows in intensity. Or vice versa, where a civilization develops on the night side, and one 'day' they explore further and begin to see a strangely persistent glow across the horizon. The eyes of that world would forever be changed by such a discovery, either way it was experienced.

The Terminator

I'm of course not talking about the movie, though my other book, *Our Cosmic Story*, does reference *The Terminator*. The term describes a visual separation between two sides of an object or set of objects. For planets in orbit around a star, the terminator is the boundary where the day side meets the night side. You can witness the effect right here on Earth. As a day passes, the Sun rises above the horizon during dawn, and after many hours, drops down again on the opposite horizon at dusk. The dawn to dusk, and dusk to dawn, transitions mark the terminator. To experience the same effect on a tidally locked world, you would need to travel many kilometers in either direction, since the planet is not rotating with respect to the parent star like it is for Earth.

Mathew C. Anderson

This static feature of a red dwarf world's landscape is what makes these planets the most interesting to think about for civilizations coping with such an effect. What a fantastical concept – a world where one-half is in perpetual daylight and the other half is in perpetual night, and right in-between is a distinct and unchanging boundary. How would a civilization survive along a terminator, especially if caught between two likely very different cultures on the day and night sides?

There are several sci-fi stories that explore how a civilization on such a divided world may develop. In the movie *White Dwarf* (1995), humans find an inhabited world tidally locked to its star. Colonies have developed across the planet, on both the day and night sides. The day side is a bustling place with sophisticated technologies; it enjoys perpetual sunshine, beautiful rolling hills and plentiful fields of food. The inhabitants are social and appreciative of the arts. Meanwhile, the night side is a war-torn medieval kingdom; it is blanketed with raging storms and devastating tornadoes, but it benefits from numerous mines full of rare metals that are used for war with the day side. The night side seems frustrated with its less than ideal landscape, and it is always attempting to stir up trouble for the day side. A massive wall separates the two along the planet's terminator, much like the Great Wall of China that functioned to keep the Mongolian hordes from invading.

Other examples of films and literature presenting tidally locked worlds include the movie *Star Trek: Nemesis*, a speculative documentary *What if the Earth STOPS Spinning*, and Isaac Asimov's novel *Nemesis*, among many other stories.

We don't know what the environment on red dwarf planets is really like, but if life exists at all on these worlds, it is bound to be just as interesting and different than our own. We need to continue observing these strange and mysterious worlds. What we have learned through our telescopes so far about confirmed exoplanets is that they have surprised us – every preconceived notion about what these worlds should be like compared to Earth has been turned upside down with the next discovery.

A NEW GENERATION OF TELESCOPES

To analyze the information needed from the light coming from exoplanets, extremely sensitivity telescopes and techniques must be employed. Up until 2009, we had no telescopes exclusive to planet hunting, and certainly none launched into space that were up to the task full-time. Governments and corporations didn't want to incur the expense of developing a telescope just on the slim chance that a few outlier exoplanets would be found. There were crude means of detecting worlds with existing telescopes back then, so we first used those to make initial discoveries that then justified new equipment.

Once it was clear that exoplanets existed in abundance, the Kepler Space Telescope was developed in the late 1990s and launched in 2009. What Kepler has since discovered has delighted scientists and the public. The exciting news out of the data collected so far is that just about every star seems to host at least one planet. This includes multi-star systems, as well as the smaller red dwarfs we've been talking about. As detection techniques keep getting refined and new telescopes are built, we're discovering ever smaller planets than even Mars and, most importantly, at orbital distances where liquid water may flow on the planet's surface.

There are some very exciting telescopes being built over the next couple of decades, many of which are dedicated entirely to planet hunting. The most sensitive will be launched into space. Two prominent upcoming ones will replace existing telescopes with state-of-the-art instruments: The Transiting Exoplanet Survey (TESS), and the James Webb Space Telescope (JWST).

TESS is due to launch in early 2018 (by the time you read this book) and is to be the successor of Kepler. It will also use the transit photometry method of detection. This method probes stars for planets that pass in front of their light. The benefit of this method is that we can understand in great detail a planet's overall size, mass, water content, and even some atmospheric properties. The downside of this method, which both Kepler and TESS use, is that only about 1% of all star systems will be aligned just right for us to detect orbiting planets. It's also not quite as good at detecting an atmosphere as we would get with a direct imaging technique. Still, this will be sufficient to establish some statistical averages. TESS will be used to build a survey of systems that JWST will later follow up on.

The successor to the Hubble Space Telescope, JWST is scheduled to launch in 2020 (or later, if it keeps getting delayed). Much like Hubble, JWST will be tasked with other research priorities, not just planet hunting. The telescope will be the most advanced we've ever sent into space, and the costliest at more than 10 billion USD. At several times Hubble's size, it will have to be folded up for launch and then unfurled in space. As you can imagine, the process of unfurling is complicated and carries many risks, so ongoing testing is being done to get it right. Once in space, if the telescope malfunctions, it will be so far from Earth at 1.5 million kilometers that it will be unrepairable without costly missions that take years to complete.

JWST will be observing objects in the infrared part of the electromagnetic spectrum. You'll recall that this is the part red dwarf stars emit most of their energy. JWST will therefore need to be far from the Sun, otherwise it will pick up the Sun's energy, drowning out any distant signals. The telescope itself will have to be extremely cold, otherwise its own electronic heat will distort the image; in fact, it will be operating at -225 °C (373 °F) below zero! JWST carries a cryocooler to get the temperature this low, as surrounding space itself is much warmer.

The best location for JWST is what's called a Lagrange point. These are points in space between two large bodies, like the Sun and Earth, where the gravitational pull is balanced. Place an object there and it will stay there for a very long time, instead of being pulled more strongly one way or another. There are five such gravitationally stable points around any two large bodies. Only one Lagrange point for Earth, L2, will be a suitable location for JWST. L2 is located on the far side of the planet from the Sun, beyond the moon's orbit.

Habitable Exoplanets: Red Dwarf Systems Like TRAPPIST-1

There are even more sensitive telescopes planned, including a few that will be built on the ground, such as the W. M. Keck in Hawaii and the Atacama Large Millimeter/submillimeter Array in Chile. What makes these telescopes possible are three primary technological advances achieved over the past decade. The first is the basic idea that more collecting area is better. The larger the mirror, or the ability to combine multiple mirrors into one virtual telescope, provides the power needed to view tiny planets light years away. The second technology is what's called adaptive optics, where a telescope deforms the mirror (or mirrors) to instantly compensate for the fluctuations in the atmosphere that normally distort incoming light. Finally, there is the processing hardware itself that includes high bandwidth Internet lines that send incoming data to other observatories for follow-up analysis.

Looking back at telescopes placed in space, there is also the intriguing direct planet observation technology, and many more cutting-edge observatories that are just now being thought up. The Starshade is my favorite concept of the direct observation technique. In development by Northrop Grumman, the concept is remarkably simple. Two parts would be sent into space, one being the Starshade itself and the other being the telescope. As its name implies, the Starshade is a very large sheet of material that blocks the light from the target star. It is sent many kilometers beyond the telescope so that orbiting planets of the target star can be viewed just beyond the shade's edges.

The name of the Starshade came about due to its unique design, which looks very much like the arrangement of petals on a sunflower. This special design was necessary to overcome the diffraction of light bending around the Starshade's edges. If the Starshade was simply a circular pattern, then what you would see in the telescope is a halo of light that would block any planets from view, rendering a billion-dollar telescope useless. Instead, you have what is essentially a coronagraph that focuses the star's light to the center of the Starshade, freeing the telescopes view for observing surrounding planets. Direct observing technologies are several years off yet. By the time they arrive though, we'll have a significant understanding of which star systems to point these telescopes toward!

I would like to thank Fraser Cain, founder and owner of the Universe Today online astronomy website, for developing the following informative two-part video series that describes more about these telescopes: https://youtu.be/XR_uEh0qhk4

SUMMARY: A UNIVERSE FILLED WITH HABITABLE EXOPLANETS

For most of my books, I prefer to write a more concise big picture view of the topic, instead of drilling down into specifics that distract from that big picture view. The goal of this book was to introduce to you the idea of red dwarf planets and what the current understanding is for the chance of life and civilization on these worlds. For perspective, I also gave you an overview of the habitable conditions right here on Earth that hint at the likely boundaries of habitability on other worlds. We then ended with the intriguing concept of tidally locked planets and what this feature and the terminator boundary means for civilization.

There are no doubt serious challenges of life's ability to evolve and survive on exotic red dwarf planets. While Earth has a protective magnetic field, an atmosphere, and water of the right balance to allow life to thrive, this will not be the case on most (or perhaps any) red dwarf planets. However, if we can confirm the suspected water inventories on the TRAPPIST-1 worlds and in other planetary systems, then this will be an important step to confirming life exists elsewhere in the Universe. The latest research builds an overall promising view of this possibility, but we need to back that view up with actual evidence.

There are countless stars in the Universe that are long-lived enough to support habitable planets. We are just beginning the search and in the next couple of decades we will surely see discoveries that confirm our theories. Thousands of planets have already been discovered, so there's cause for hope that one day we will not only discover a habitable world around another star, but one that is in fact inhabited. If we can accomplish this at least once, we will know that life is abundant across the Universe, and is quite possibly more fantastical than our science fiction stories could dream up. It may even be found someday a world that has what's called 'super-habitable' conditions where life thrives in even greater abundance than it does on Earth.

As Carl Sagan once said, *"Absence of evidence is not evidence of absence!"*

Habitable Exoplanets: Red Dwarf Systems Like TRAPPIST-1

A Quick Checklist on Red Dwarf Exoplanet Habitability
- Is the star past its early years of active flaring?
- Does the planet have a magnetic field of any kind?
- Does the planet have a thick enough atmosphere?
- Is there any kind of atmospheric recycling occurring?
- Does the planet have liquid water on its surface?
- And for the support of a technological civilization, does the planet have between 10-40% land coverage?

Most Promising Exoplanets
(From the Earth Similarity Index (ESI) where 1.0 is deemed at least as habitable as Earth. For reference, Mars is at 0.797.)
- Gliese 581 g (0.89)
- Kepler-438b (0.88)
- Ross 128 b (0.86)
- Luyten b (0.86)
- Gliese-3323b (0.86)
- Proxima Cen b (0.85)
- TRAPPIST-1e (0.85)
- GJ 667 Cc (0.84)
- Kepler-442 b (0.84)
- Kepler-62e (0.83)
- Gliese 832 c (0.81)

As of February 2018, there are at least 600 multiplanetary systems confirmed with a total of over 2,750 planets in these and single planetary systems. Over 90% of detections have come from the transit method.

Given all that we have covered in this book, if I were to sum up the chances of finding complex life around a red dwarf star, a star that's at least a couple of billions of years old, I would personally rate the chance at 1 of 5 (20%). This percentage is still nearly twice as many stars as the total of all Sun-like stars in our galaxy. By this statistic alone, red dwarf worlds demand all the telescope time they can get.

SUGGESTIONS FOR FURTHER READING

There are many topics in this book that draw upon knowledge and insights from many scientists and great thinkers. Here is a collection of books that will give you an even bigger picture view of our place in the cosmos, and the chance of one day discovering an inhabited exoplanet:

Baggott, Jim. *Origins*. Oxford University Press, 2015.
Carroll, Michael. *Earths of Distant Suns*, 2017.
Davies, Paul. *Are We Alone?*. UK: Penguin Books, 2015.
Davies, Paul. *The Eerie Silence*. UK: Penguin Books
Fichman, Frederick. *The SETI Trilogy*. Frederick Fichman, 2014.
Hawking, Stephen. *A Briefer History of Time*. NY: Bantam Dell, 2005.
Krauss, M. Lawrence. *The Greatest Story Ever Told...* Atria Books, 2017.
Nolan, Christopher. *The Science of Interstellar*. W.W. Norton & Comp., 2014.
Osgood, Don. SETI 2020: *A Roadmap for the Search for Extraterrestrial Intelligence*. Amazon.com, 2002.
Sagan, Carl. *Contact*. Pocket Books, 1998.
Sagan, Carl. *Cosmos*. Ballatine Books, (Multiple Years).
Savage, Marshall T. *The Millennial Project*. Empyrean Pub, 1993.
Stevenson, David S. *Under a Crimson Sun*. Springer, 2013.
Summers, Michael. *Exoplanets*. Smithsonian Books, 2017.
Tasker, Elizabeth. *The Planet Factory*. Bloomsbury Sigma, 2017.
Ward, Peter and Brownlee, Donald E., *Rare Earth: Why Complex Life Is Uncommon in the Universe*, 2000.
Webb, Stephen. *If the Universe is Teeming with Aliens...*, 2015.
Weir, Andy. *The Martian*. Random House, 2014.
Zubrin, Robert. *Entering Space...* TarcherPerigee, 2000.
Zubrin, Robert. *The Case For Mars...* Free Press, 2011.

EMBEDDED REFERENCES

Habitable Exoplanets
1. https://bit.ly/2Ik5xvY
2. https://go.nasa.gov/2CmQAL1
3. http://exoplanet.eu/catalog/

Introducing Red Dwarf Stars
1. https://bit.ly/2wz6ej7
2. https://bit.ly/2IkpG50

Surprise! Red Dwarfs Host Planets Too
1. https://bit.ly/2zYP2Rh
2. https://bit.ly/2mRABdd
3. https://go.nasa.gov/1J7Zje4
4. http://www.trappist.one/
5. https://bit.ly/2H6sGCM

A Red Dwarf Planetary System
1. https://keplerscience.arc.nasa.gov/
2. https://go.nasa.gov/2Ezln58
3. http://universesandbox.com/
4. https://bit.ly/2G836Mi

A Star's Temperament
1. https://go.nasa.gov/2rKw3HX
2. https://bit.ly/2pVRjKO
3. https://go.nasa.gov/2oydf00
4. https://bit.ly/2vMnaxL
5. https://bit.ly/2uD09C7
6. https://bit.ly/2Im2MKm

A Planet's Shielding
1. https://bit.ly/2H2rVd3
2. https://bit.ly/1IBswuK
3. https://go.nasa.gov/2pZq7Kn
4. https://bit.ly/2pVA37O
5. https://bit.ly/2EcdnFh
6. https://bit.ly/2H3gjXQ
7. https://bit.ly/2CAZeTd
8. https://bit.ly/2Gpp15L
9. https://go.nasa.gov/28P2uML
10. https://bit.ly/2DHbyk3

Water Worlds
1. https://bit.ly/2ecxlco
2. https://go.nasa.gov/2ImYAKF
3. https://go.nasa.gov/2p9BWzh
4. https://bit.ly/2GPGW4S
5. https://bit.ly/2pXX3TP
6. https://go.nasa.gov/2JN8WoA

A Delicate Balance
1. https://bit.ly/2E9tsvr
2. https://bit.ly/2uEju5L
3. https://bit.ly/2EbN0Q2
4. https://bit.ly/2Ik7qIW
5. https://bit.ly/2pW2F1A
6. https://go.nature.com/2pVCSpq

Introducing Tidally Locked Worlds
1. https://bit.ly/2CB5fPE
2. https://bit.ly/2ecxlco
3. https://bit.ly/2Jdu4UQ
4. https://bit.ly/2pVpvpR
5. https://bit.ly/2JcbDQ7

TERM REFERENCES

51 Pegasi b: https://go.nasa.gov/2CmQAL1
Alexander Wilszczan: https://bit.ly/2H0tg8c
Allan Telescope Array: http://bit.ly/1S4UtCb
Andrew Grey's discovery: http://bit.ly/2CjnPPg
Artificial Intelligence: http://bit.ly/1R7lUu8
Atacama Large Millimeter Array (ATMA): http://bit.ly/2zXnNH0
Atmospheres: http://bit.ly/2q6SCJm
Atmospheric Cold Trap: http://bit.ly/2C6DBZV
Carbon dioxide (CO2) (as a greenhouse gas): http://bit.ly/2DKt46O
Celestron CGX-L 1400: http://bit.ly/2lzBeIJ
Collision energy of space debris: http://bit.ly/1O1y3D4
CO2 concentrations in Earth's past: http://bit.ly/2zyvOq2
Desiccated planets: https://www.nature.com/articles/504221a
Direct Imaging technologies: http://bit.ly/2zXnj3s
Drake Equation for kids: http://bit.ly/1cpiQ6Q
Dynamo Theory: https://bit.ly/2uVUiYD
Earth Similarity Index: http://bit.ly/1nVCJrl
Exoplanet overview: http://bit.ly/2EsEoWz
Frank Drake history with Enrico Fermi: http://bit.ly/1WcD93Z
Gliese 876 b: http://bit.ly/2zYP2Rh
Great Wall of China: http://bit.ly/1wAzz4z
Habitable zones: http://bit.ly/2C7TVJE
James Webb Space Telescope: https://www.jwst.nasa.gov/
Lagrange Points: http://bit.ly/2ueSyYP
Lithiophosphate: http://bit.ly/2Cu9hvC
Kepler-90 system: https://apod.nasa.gov/apod/ap171218.html
Kepler Space Telescope: https://go.nasa.gov/2C5XrV2
Mars: http://bit.ly/2vir4zw
Mercury: https://solarsystem.nasa.gov/planets/mercury
Moist Greenhouse: https://bit.ly/2HhUPqh
Nemesis novel by Isaac Asimov: http://bit.ly/2lxrLSi
Oregon Trail: http://bit.ly/1p7oPLH
Phosphammite: http://bit.ly/2CnzFbm
PlanetHunters.org: https://www.planethunters.org/
Planet size and gravity on atmospheric gases: http://bit.ly/2H4PxxY
Planetary magnetic fields: http://bit.ly/2zZlDGA
Planetary migration: https://go.nasa.gov/1rRTkX0
Plate tectonics: http://bit.ly/2pn3zVm
Proxima Centauri: http://bit.ly/2Czt3U1
Radio communications travel distance: http://bit.ly/13onzcd
Red Dwarf star population: http://bit.ly/2lCbkDl

Ross 128 planet (Discovered in 2017): http://theatln.tc/2A0FXfd
Search for exoplanet life: http://bit.ly/1ySMmNn
Size of Earth comparisons: http://bit.ly/1NuKTXl
Snowball State: https://bit.ly/2qk3HVC
Space resources: http://bit.ly/1MkNGoU
Spherical Minimum: http://bit.ly/2j2tv7D
Star Trek series: http://www.startrek.com/
Star Trek: The Next Generation: http://bit.ly/1ncvVOD
Super-Earth range of potential: http://bit.ly/IPMmvo
Super-Earth comparison to Earth: http://bit.ly/IPMmvo
Super-Earth surface gravity: http://bit.ly/1g1jqjh
Supernova rates: http://bit.ly/1OPPPcG
Supervolcanoes: https://on.doi.gov/2uUGKwK
Terminator on a Tidally Locked World: https://bit.ly/2CB5fPE
Tidally locked worlds: http://bit.ly/2CB5fPE
Transiting Exoplanet Survey Satellite (TESS): https://tess.gsfc.nasa.gov/
TRAPPIST-1: http://www.trappist.one/
Travel to other star systems: http://bit.ly/1Rw1meP
Types of civilizations: http://bit.ly/1MNhODO
Types of Ultraviolet radiation: http://stanford.io/2z8sctZ
Ultraviolet radiation overview: https://go.nasa.gov/2plu9zw
Universal Sandbox game: http://universesandbox.com/
Universe Today (website): https://www.universetoday.com/
Venus: http://bit.ly/2ycnQxZ
Volcanism: https://www.britannica.com/science/volcanism
W.M. Keck Observatory: http://www.keckobservatory.org/
Water (Ocean) worlds in our solar system: https://go.nasa.gov/2p9BWzh
Water world potential for life: http://bit.ly/2tzbRJM
White Dwarf movie: http://imdb.to/2CBSLXO
Zooniverse: https://www.zooniverse.org/

SCIENTIFIC PAPERS

- **Feb. 9, 2018** – Leaky atmosphere linked to lightweight planet: http://bit.ly/2H4PxxY
- **Feb. 6, 2018** – Inward Migration of the TRAPPIST-1 Planets as Inferred From Their Water-Rich Compositions: https://bit.ly/2G836Mi
- **Jan. 31, 2018** – The Nature of the TRAPPIST-1 Exoplanets: http://bit.ly/2BknY3c
- **Jan. 10, 2018** – The Moist Greenhouse is Sensitive to Stratospheric Temperature: https://bit.ly/2uEju5L
- **Dec. 9, 2017** – Atmospheric escape from the TRAPPIST-1 planets and implications for habitability: http://bit.ly/2reH5qP
- **Nov. 29, 2017** – Stratosphere circulation on tidally locked ExoEarths: https://bit.ly/2Ev9tYo
- **Oct. 17, 2017** – NIR-driven Moist Upper Atmospheres of Synchronously Rotating Temperate Terrestrial Exoplanets: http://bit.ly/2CmRCqe
- **Sep. 12, 2017** – The Dehydration of Water Worlds via Atmospheric Losses: http://bit.ly/2zYhEtR
- **Aug. 16, 2017** – Solar Wind Interaction and Impact on the Venus Atmosphere: http://bit.ly/2lAzpua
- **Aug. 2, 2017** – The Surface UV Environment on Planets Orbiting M-Dwarfs: Implications for Prebiotic Chemistry & Need for Experimental Follow-up: http://bit.ly/2CAZeTd
- **Jul. 31, 2017** – Temporal Evolution of the High-Energy Irradiation and Water Content of TRAPPIST-1 Exoplanets: http://bit.ly/2CmRzLl
- **Jul. 10, 2017** – Physical Constraints on the Likelihood of Life on Exoplanets: http://bit.ly/2CkUa8q
- **Jun. 14, 2017** – The Threatening Magnetic and Plasma Environment of the TRAPPIST-1 Planets: http://bit.ly/2uO0d0W
- **Mar. 16, 2017** – Assessing the Habitability of the TRAPPIST-1 System Using a 3D Climate Model: http://bit.ly/2C684at
- **Mar. 4, 2017** – Seven Temperate Terrestrial Planets Around the Nearby Ultracool Dwarf Star TRAPPIST-1: http://bit.ly/2CyqAsP
- **Feb. 27, 2017** – The Habitability of Planets Orbiting M-dwarf Stars: http://bit.ly/2ecxlco
- **Feb. 27, 2017** – The (Potentially) Habitable Worlds of TRAPPIST-1: http://bit.ly/2DK2Xx9
- **Dec. 16, 2016** – Evidence for Atmospheric Cold-trap Processes in the Noninverted Emission Spectrum of Kepler-13Ab: http://bit.ly/2zYX0Kn
- **Jun. 6, 2016** – A Combined Transmission Spectrum of the Earth-sized Exoplanets TRAPPIST-1 b and c: http://bit.ly/2C8lZN9

Mathew C. Anderson

- **Nov. 2, 2015** – Influence of the UV Environment on the Synthesis of Prebiotic Molecules: http://bit.ly/2CpRC8i
- **Sep. 24, 2015** – Tidal heating of Earth-like exoplanets around M stars: Thermal, magnetic, and orbital evolutions: https://bit.ly/2JcbDQ7
- **Oct. 28, 2014** – Kepler Flares I. Active and Inactive M Dwarfs: http://bit.ly/2DKNVHn
- **Jan. 14, 2014** – Role of ocean heat transport in climates of tidally locked exoplanets around M dwarf stars: https://bit.ly/2Jdu4UQ
- **Oct. 9, 2013** – High Stellar FUV/NUV Ratio and Oxygen Contents in the Atmospheres of Potentially Habitable Planets: http://bit.ly/2qeoj3Q
- **Jan. 8, 2012** – The Runaway Greenhouse: Implications for Future Climate Change, Geoengineering and Planetary Atmospheres: http://bit.ly/2DHbyk3
- **Oct 28, 2010** – The evolution of the marine phosphate reservoir: https://go.nature.com/2GZz1zD
- **May 31, 2010** – The Effect of a Strong Stellar Flare on the Atmospheric Chemistry of an Earth-like Planet Orbiting an M dwarf: http://bit.ly/2DKChvR

ABOUT THE AUTHOR

Mathew has been exploring the boundaries of what it means to be human since voluntarily stepping in wet cement while on the way to his first kindergarten class. The adventures and lessons learned since then have only become more unexpected and profound.

Mathew studies physics, astronomy, and related areas, and often consults with various production studios on the topics found in this book. His goal is to raise everyone's awareness on why our existence as a civilization is so precious. Mathew is also a senior community manager in the computer game industry.

For more writings by the author:
www.amazon.com/author/mathewanderson

You can also follow Mathew's work on Twitter:
@OurCosmicStory and @MathewAnderson

Printed by Amazon Italia Logistica S.r.l.
Torrazza Piemonte (TO), Italy